Funtastic Math Games

by Debbie Haver,
Alice Koziol,
Elaine Haven,
and Dan Mulligan

illustrated by Milton Hall

cover by Margo DePaulis

Publisher
Instructional Fair • TS Denison
Grand Rapids, Michigan 49544

Instructional Fair • TS Denison grants the individual purchaser permission to reproduce patterns and student activity materials in this book for noncommercial individual or classroom use only. Reproduction for an entire school or school system is strictly prohibited. No other part of this publication may be reproduced in whole or in part. No part of this publication may be reproduced for storage in a retrieval system, or transmitted in any form or by any means, electronic, mechanical, recording, or otherwise, without the prior written permission of the publisher. For information regarding permission, write to Instructional Fair • TS Denison, P.O. Box 1650, Grand Rapids, MI 49501.

ISBN: 1-56822-747-7
Funtastic Math Games
Copyright © 1998 by Instructional Fair • TS Denison
2400 Turner Avenue NW
Grand Rapids, Michigan 49544

All Rights Reserved • Printed in the USA

Table of Contents

A Perfect Square Match 1
A Step Back in Time 4
Billion-Dollar Hot Dogs 8
Competing Integers 12
Competing Percentages 14
Cover the Area 16
Cross-Country Trip 18
Decimal Dominoes 21
Desperately Seeking Decimals 24
Eradication 26
Find Me If You Can 29
Fraction Fish 31
Geoboard Match 35
Geometric Mystery 38
Math Monopoly 41
Math Scavenger Hunt 44
Metric Match 46
Mystery Proportion 48
Name That Symbol 50
Number Mystery 55
Order War 56
Percent Match 58
Positive or Negative Numbers 60
Rational Madness 61
Shapes, Shapes, Shapes 63
Spin-Roll Expressions 66
Square T^2 72
Stars and Pounds 74
Strategic Plotting 77
Stem-and-Leaf Logic 79
Sweet Success 81
Twins 83
Westward Go! 86

About This Book

The mathematics community is encouraged to teach using a variety of strategies. This book offers a wide variety of games—card games, board games, dice games, and word and picture games to reinforce mathematical skills and to develop mathematical power in students. Games are naturally fun and motivating; therefore, using games as an instructional strategy will make mathematics more inviting for your students.

The games are designed for you to use materials available in your school. Most games provide the necessary materials for duplication. We suggest that you laminate these materials to ensure their durability. The first page of each game is written in a brief outline form allowing you a quick glance at the objective, materials needed, and the number of players. The directions may be copied and given to students.

We recommend that prior to starting any game you do the following:

- Explain the purpose of the activity, for it is essential that students make the connection to mathematics.

- Play the game with student volunteers to allow the class to observe the process and ask questions about the procedures or game rules.

- Debrief the game. Direct students to explain the purpose of the game; write what mathematics the game helped them learn and how the game helped them learn it; or tell their partner what they liked or disliked about the game.

Most games can be adapted to include other mathematical skills. You may wish to construct addition game cards that meet your specific instructional objective. ENJOY!

A Perfect Square Match

Objective: to compute the product of perfect square binomials

Materials needed: 24 cards cut out and laminated

Number of players: two players

Directions:

1. Shuffle the cards. Arrange the cards face down in a rectangle.

2. The first player turns over two cards. If a card contains a perfect square binomial, determine its product to see if it matches the other card. If the two cards represent equivalent expressions, the player scores one point and turns over two more cards. If they don't match, the player turns the cards face down again, no points are scored, and it becomes the next player's turn.

3. Players take turns until all cards are matched. The player with the most points wins.

$(x-1)^2$	$x^2 - 2x + 1$	$(x+1)^2$
$x^2 + 2x + 1$	$(2x-3)^2$	$4x^2 - 12x + 9$
$(2x+3)^2$	$4x^2 + 12x + 9$	$(x-2)^2$
$x^2 - 4x + 4$	$(x+2)^2$	$x^2 + 4x + 4$

$(2x - 1)^2$	$4x^2 - 4x + 1$	$(2x + 1)^2$
$4x^2 + 4x + 1$	$(x - 3)^2$	$x^2 - 6x + 9$
$(x + 3)^2$	$x^2 + 6x + 9$	$(2x - 2)^2$
$4x^2 - 8x + 4$	$(2x + 2)^2$	$4x^2 + 8x + 4$

A Step Back in Time

Objective: to provide practice for divisibility

Materials needed: "A Step Back in Time" gameboard, a spinner numbered 1-9, a token for each player

Number of players: two to four players

Directions:

1. Each player selects a token and places his or her token at START on the "A Step Back in Time" gameboard.
2. The player whose last name begins with the letter closest to the end of the alphabet is first.
3. The first player spins the spinner and looks for the first number on the gameboard that can be evenly divided by the number on the spinner.
4. The player then identifies that number on the "footprint" on the gameboard and then moves his or her token to that position.
5. Each player follows in turn spinning the spinner and then moving his or her token to the first number on the gameboard that can be divided evenly by the number on the spinner.
6. If a player makes an error, that player may be challenged by another player. The player choosing an incorrect answer is not allowed to move and thus loses a turn.
7. If a player lands on a footprint with a message, the player must follow the directions given.
8. The winner of the game is the first player to reach the FINISH.

A Step Back in Time

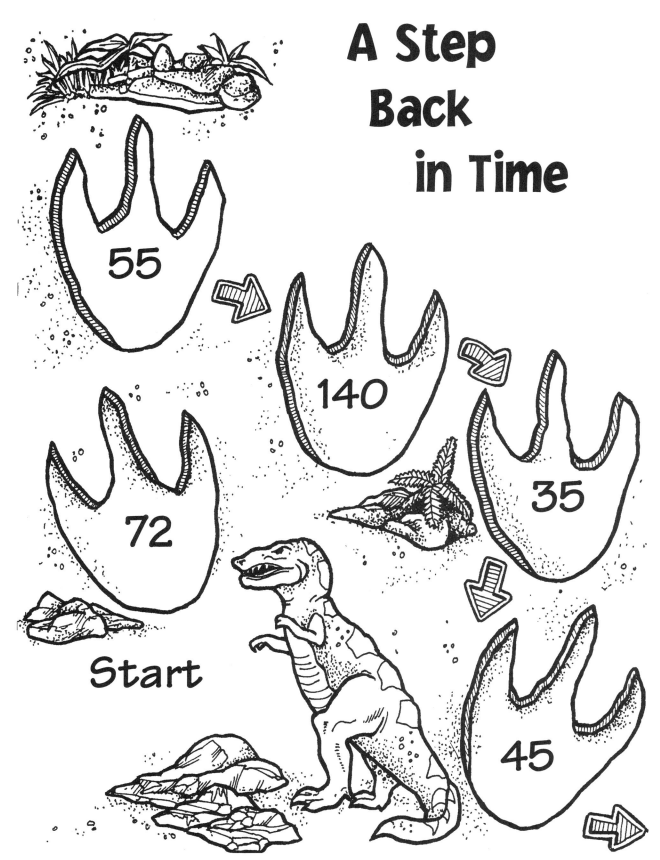

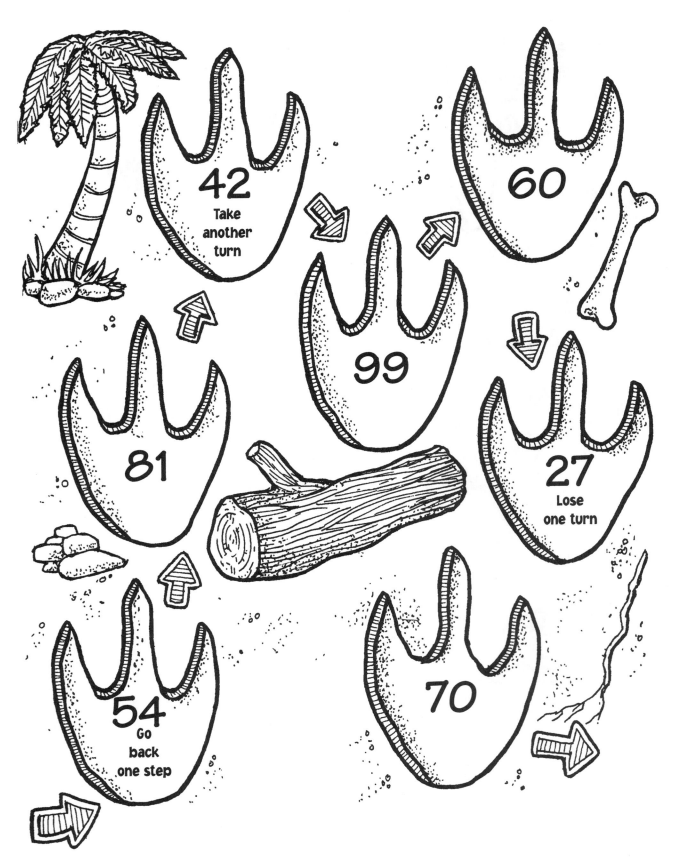

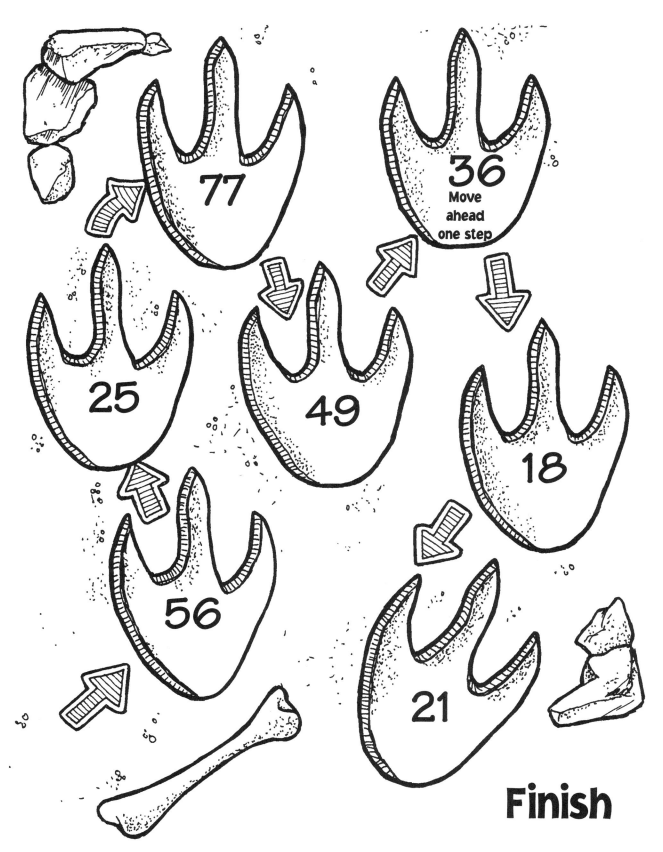

Billion-Dollar Hot Dogs

Objective:
to provide practice for place value

Materials needed:
four sets of ten clue cards printed with questions concerning place values up to and inclusive of one billion
four sets of the "Billion-Dollar Hot Dog Pieces"

number of players: four players

Directions:

1. The object of the game is to arrange all the pieces of the hot dog in correct place value order.

2. Each player takes a sheet of the "Billion-Dollar Hot Dog Pieces" to cut apart.

3. The cards are shuffled and placed face up in the middle of the playing area.

4. The oldest player begins the game by taking the top card from the pile and answering the question. The player then checks the back of the card for the correct answer. If the answer is correct, the player finds the corresponding piece among his or her billion-dollar hot dog pieces and begins to place the hot dog pieces in correct order.

5. If the player is unable to answer the question correctly, he or she is not allowed to add a piece of the puzzle part to the billion-dollar hot dog.

6. The winner is the first player to get all ten pieces in the correct place value order.

Clue Cards

1. The ones place is an even number greater than five and less than seven.	2. The tens place is two ninths of eighteen.	3. The hundreds place is two to the third power.
4. The thousands place is the third odd number.	5. The ten thousands place is the quotient of 49 and 7.	6. The hundred thousands place is the product of three and three.
7. The millions place is equivalent to 25% of 8.	8. The ten millions place is the second prime number.	9. The hundred millions place is the difference between the product of four times five and two times ten.
	10. The billions place is equal to four times .25.	

800	40	6
900,000	70,000	5,000
000,000,000	30,000,000	2,000,000

| 1,000,000,000 |

Billion-Dollar Hot Dog Pieces

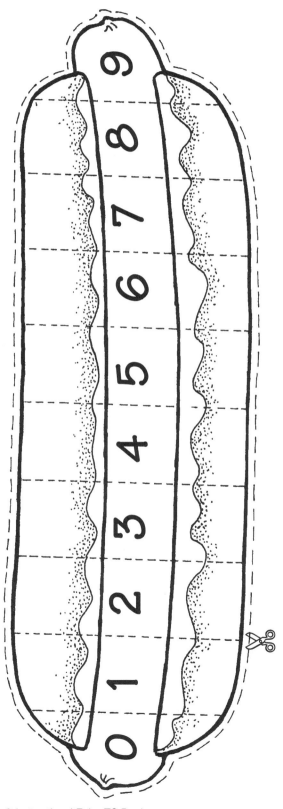
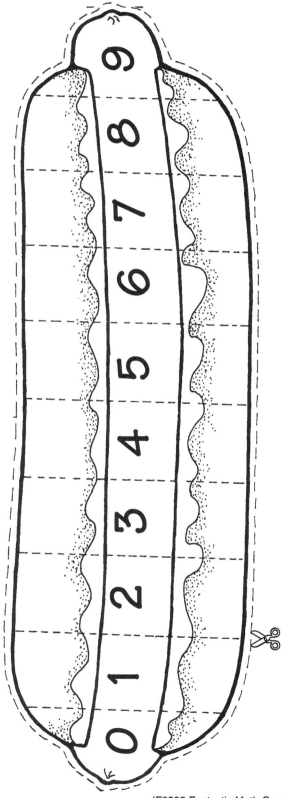

Competing Integers

Objective: to practice computing integers mentally

Materials needed: chalkboard

number of players: whole class

Directions:

1. Divide the class into two teams. Place each student's name from each team on an index card. Keep the two decks separate. Shuffle the name cards.

2. Take a name card from each stack and call the students' names. The two students go to the chalkboard and wait for you to call out a problem.

3. Once you state the problem, each student is to write the problem on the board with the answer. The first student to correctly write and then solve the problem wins a point for his or her team.

4. After every member of each team has an opportunity to participate, the team with the greatest number of points wins.

Problems

-28 + 24	-4	(-9)(5)	-45
-2 + (-3)	-5	(45)(-1)	-45
14 + 8	22	(5)(-1)(-5)	25
-16 - (-4)	-12	4 + 26	30
(-6)(-8)	48	-12 - (-6)	-6
-63 divided by -9	7	-6 - 25	-31
-45 + 3	-42	5(-2)8	-80
-32 divided by 4	-8	(-6)(-11)	66
12 + (-32)	-20	-64 divided by -8	8
-5 + 16	11	-6 divided by -2	3
77 divided by 7	11	15 - (-5)	20
(-2)(8)(0)	0	4(-5)	-20
-13 + 13	0	3(-3)(6)	-54
(-2)(-4)(-2)	-16	48 - (-28)	76
35 divided by -5	-7	(-3)(7)(0)	0

Create additional problems to extend the game.

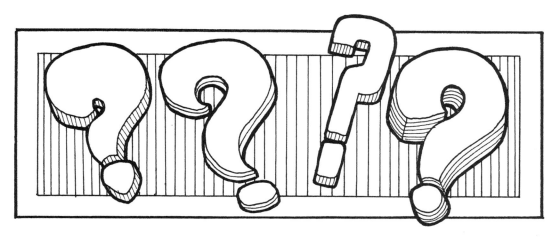

Competing Percentages

Objective: to practice solving percentage problems

Materials needed: chalkboard, paper and pencil, calculator

number of players: whole class

Directions:

1. Divide the class into two teams. Place each student's name from each team on an index card. Keep the two decks separate. Shuffle the name cards. Place two desks in front of the room.

2. Take a name card from each stack and call the students' names. The two students go to the desks in front of the room and wait for you to call out a problem.

3. Once you state the problem, each student is to solve the problem on paper and then write the problem on the board with the answer. The first student to correctly explain and legibly write the problem wins a point for his or her team.

4. After every member of each team has an opportunity to participate, the team with the greatest number of points wins.

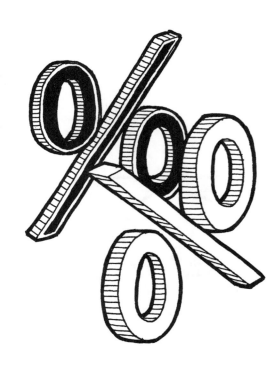

Problems

1. What percent of 20 is 16?

   80%

2. 24 is what percent of 25?

   96%

3. 10% of what number is 42?

   420

4. What number is 40% of 236?

   94.4

5. 50% of what number is 528?

   1,056

6. What percent of 200 is 68?

   34%

7. 3.5 is what percent of 50?

   7%

8. 30 is 60% of what number?

   50

9. What number is 2.5% of 9,600?

   240

10. 100% of 456 is what number?

    456

11. 75% of what number is 60?

    80

12. 250% of what number is 50?

    20

13. 18 is what percent of 18?

    100%

14. What percent of 66 is 44?

    66⅔%

15. What number is 37½% of 120?

    45

16. What percent of 48 is 2.4?

    5%

Create additional problems to extend the game.

Cover the Area

Objective: to develop the concept of area

Materials needed:
2 one-hundred grids
number cubes
crayons or colored pencils

Number of players: two players

Directions:

1. Each player rolls one number cube. The player with the highest number starts the game.

2. The first player rolls both number cubes to determine the length and width of his or her rectangle. Using one color, the first player shades the area on his/her one-hundred grid.

3. The next player rolls the number cubes and shades the area on his or her one-hundred grid.

4. On the next turn, the first player chooses another color to shade in the area determined by the roll of the number cubes.

5. With each turn, the players must choose another color to shade the area determined by their roll.

6. Play continues until one player's grid is completely shaded. That player is the winner.

Cross-Country Trip

Objective: to review basic multiplication facts by playing a game in which students take a trip from Virginia to California

Materials needed:
one gameboard
small markers
spinner 1-9 or two sets of number cards 1-9

number of players: two to four players

Directions:

1. Students play this game in groups. The first player spins the spinner twice to get two numbers to multiply or chooses two number cards. If he or she names the product correctly, and the product is in the first circle, that student gets to move to that state.

2. If the student has a product that is not in the next state, or answers incorrectly, he or she stays there until the next turn.

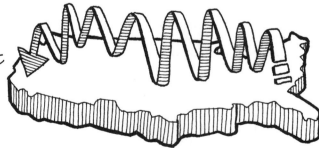

3. The student who gets to California first is the winner.

1	2	3
4	5	6
7	8	9

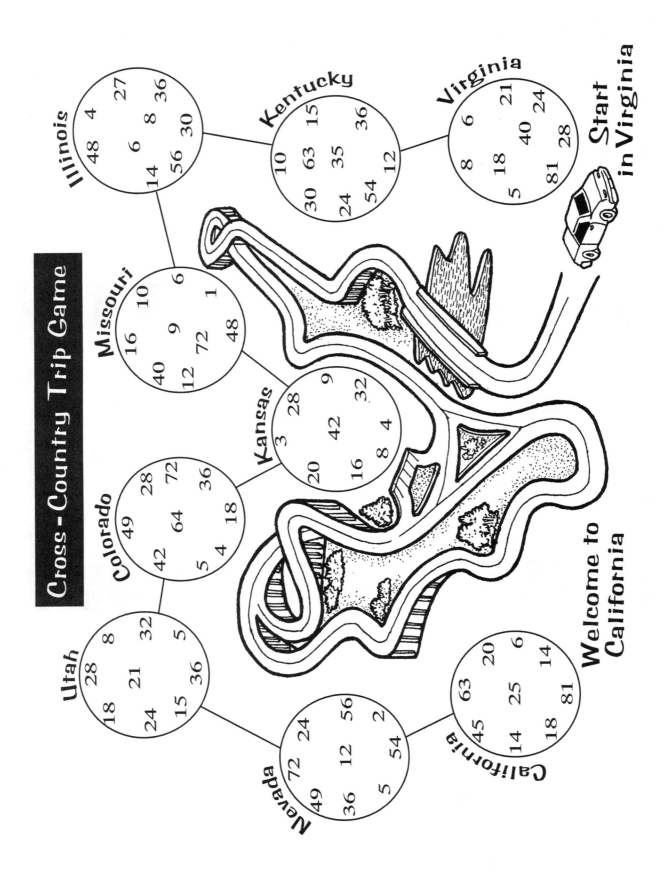

Decimal Dominoes

Objective: to match equivalent fraction and decimal dominoes

Materials needed: dominoes duplicated on cardstock

Number of players: two players

Directions:

1. Each player chooses five dominoes, keeping the numbers hidden.

2. The first player lays one domino face up. The opponent has to match either the fraction or the decimal with an equivalent number. Players may play on either end. If a player cannot make a match, that player must pick up another domino from the pile on the table.

3. The first player to be out of dominoes is the winner.

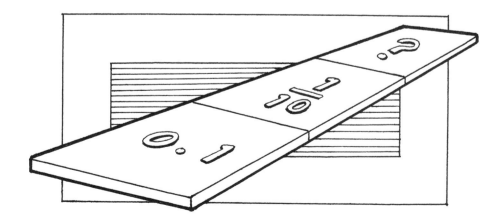

$\dfrac{1}{10}$	$\dfrac{3}{20}$	$\dfrac{6}{10}$	$\dfrac{2}{15}$	$\dfrac{1}{20}$	$\dfrac{1}{4}$
$0.08\overline{3}$	0.3	0.875	0.4	$0.1\overline{6}$	0.6
$\dfrac{2}{3}$	$\dfrac{4}{5}$	$\dfrac{1}{5}$	$\dfrac{3}{8}$	$\dfrac{1}{9}$	$\dfrac{3}{4}$
0.66	0.15	0.8	0.5	$0.\overline{3}$	$0.\overline{13}$

$\frac{7}{8}$	$\frac{1}{6}$	$\frac{2}{5}$	$\frac{1}{12}$	$\frac{1}{3}$	$\frac{4}{8}$
0.125	0.1	0.25	0.05	$0.\overline{1}$	0.375

$\frac{1}{8}$	$\frac{3}{10}$
0.2	0.75

Desperately Seeking Decimals

Objective: to provide practice for converting fractions to decimals

Materials needed:
two cubes numbered one through six
a decimal table score sheet for each player
a pencil for each player

Number of players: two to four players

Directions:

1. Each player rolls one of the number cubes to determine who starts the game. The player rolling the highest number goes first.

2. The first player rolls the two cubes and then creates a proper or an improper fraction from the numbers on the faces of the cubes.

3. The same player then converts the fraction to a decimal and records that answer on the score sheet.

4. Each player then follows in turn by rolling the cubes and recording the answer in decimal form on his or her score sheet.

5. If a player rolls the cubes and has already used the appropriate squares for those numbers, that player forfeits.

6. The winner is the first player to correctly complete a row, column, or diagonal of six squares.

Desperately Seeking Decimals

_____'s Chart

Decimal Table							
NUMERATOR							
D E N O M I N A T O R		1	2	3	4	5	6
	1						
	2						
	3						
	4						
	5						
	6						

Eradication

Objective:
to provide practice for multiplication of greater numbers, rounding to the greatest place value, and mental computation

Materials needed:
an Eradication score sheet blank cube the teacher numbers with the number 4 opposite the number 9, the number 5 opposite the number 8, and the number 6 opposite the letter E.
Another blank cube the teacher labels with the letter X opposite the number 9, the number 5 opposite the number 8, and the number 6 opposite the number 7.
The letters E and X on the cubes should be printed in red ink.

Number of players: two players

Directions:

1. Each player tosses the cube with the E printed on it. The player rolling the largest number begins the game Eradication.

2. The first player rolls both cubes and multiplies the two numbers on the cubes. The player then rounds that answer to the greatest place value and records that answer on the Eradication score sheet.

3. Each player follows in turn.

4. When the players again have a turn, they toss the cubes, multiply the two numbers they obtain when they roll the cubes, and round

that answer to the greatest place value. Now the player must multiply his or her current answer with the previous answer before recording the cumulative calculations on the Eradication answer sheet.

5. If a player rolls an E or an X on one cube and a number on another cube, the player will receive no score for that turn.

6. If the player rolls an E on one cube and an X on the other cube, the total score he or she has obtained thus far is eradicated. The unlucky player is not out of the game, however, but must again begin to accumulate points with his or her next turn.

7. The winner is the first player to accumulate 1,000 points or the person who has accumulated the highest number of points at the end of a designated time period.

Eradication	
Roll	
1	
2	
	Ans	
3	
	Ans	
4	
	Ans	
5	
	Ans	

Eradication	
Roll	
1	
2	
	Ans	
3	
	Ans	
4	
	Ans	
5	
	Ans	

Find Me If You Can

Objective: to increase mental math abilities

Materials needed:
1 one-hundred grid for each player
a counter or marker for each player
folders to create a barrier

number of players: whole group

Directions:

1. The teacher should make certain each player has a one-hundred grid, counter, and a folder. Tell players to stand folders on their sides to create barriers so other students won't see what number they are on.

2. The teacher chooses a number for each student. Students then place their markers on their numbers. The teacher calls out the directions slowly at first but then quickens the pace.

3. Players keep track on their one-hundred grid. Occasionally, the teacher should check the players' hundred grid for accuracy.

4. For example, the teacher says "Place your counters on number 46. Move your marker as you add or subtract. Add 11. Subtract 12. Add 5. Subtract 14." The teacher slowly increases the pace of directions to help players increase their mental math speed.

5. The players begin to see that adding 10 or more requires a move down and to the right. Subtracting 10 or more requires a move up and to the left.

1	2	3	4	5	6	7	8	9	10
11	12	13	14	15	16	17	18	19	20
21	22	23	24	25	26	27	28	29	30
31	32	33	34	35	36	37	38	39	40
41	42	43	44	45	46	47	48	49	50
51	52	53	54	55	56	57	58	59	60
61	62	63	64	65	66	67	68	69	70
71	72	73	74	75	76	77	78	79	80
81	82	83	84	85	86	87	88	89	90
91	92	93	94	95	96	97	98	99	100

Fraction Fish

Objective: to identify equivalent fractions in a game of Fish

Materials needed: a deck of fraction cards for each pair of players

number of players: two players

Directions:

1. Shuffle the cards. Each player is dealt five cards. The remaining cards are placed face down on the table.

2. Just as in the game of Fish, students take turns asking another student for a certain card. When a student gets two equivalent fractions, he or she lays both cards down.

3. If a player asks another student for a fraction and receives it, he or she continues asking and making sets until the opponent says "Fish" because he or she does not have that fraction card to give. That player draws a card from the fishing hole and the next student asks for a card.

4. The object of the game is to get the most equivalent fractions.

$\frac{1}{2}$	$\frac{6}{12}$	$\frac{18}{36}$	$\frac{4}{8}$
$\frac{1}{4}$	$\frac{3}{12}$	$\frac{4}{16}$	$\frac{6}{24}$
$\frac{3}{4}$	$\frac{6}{8}$	$\frac{9}{12}$	$\frac{12}{16}$
$\frac{1}{3}$	$\frac{2}{6}$	$\frac{3}{9}$	$\frac{4}{12}$

$\frac{2}{3}$	$\frac{8}{12}$	$\frac{24}{36}$	$\frac{6}{9}$
$\frac{1}{5}$	$\frac{2}{10}$	$\frac{4}{20}$	$\frac{5}{25}$
$\frac{1}{8}$	$\frac{3}{24}$	$\frac{2}{16}$	$\frac{6}{48}$
$\frac{5}{8}$	$\frac{10}{16}$	$\frac{15}{24}$	$\frac{20}{32}$

$\frac{7}{8}$	$\frac{14}{16}$	$\frac{35}{40}$	$\frac{28}{32}$
$\frac{1}{10}$	$\frac{3}{30}$	$\frac{4}{40}$	$\frac{5}{50}$
$\frac{1}{12}$	$\frac{3}{36}$	$\frac{6}{72}$	$\frac{5}{60}$
$\frac{4}{15}$	$\frac{8}{30}$	$\frac{12}{45}$	$\frac{16}{60}$

Geoboard Match

Objective: to use the geoboard and dot paper to communicate directions to another player behind a folder or other barrier who will try to duplicate the geoboard design on his or her dot paper

Materials needed:
geoboards
dot paper
rubber bands

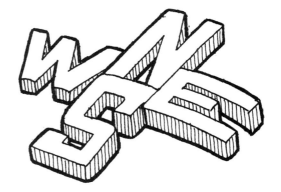

Number of players: two players

Directions:

1. On the board write directional words, such as parallel, intersecting, perpendicular, diagonal, horizontal, vertical, N, S, E, W, and intermediary directions, such as NE, SE, NW, SW, etc. Discuss each word.

2. Pairs of students work together with a barrier between them. They take turns with one student building a design on the geoboard.

3. The first student explains the design while the other student draws the design on the dot paper. For example, the first student may say, "Place one rubber band from north to south on the middle row of nails. Use another rubber band going across from east to west along the middle row of nails perpendicular to the other rubber band. The last rubber band runs parallel to the first rubber band north to south along the last row of nails to the right." (See diagram at the top of the next page.)

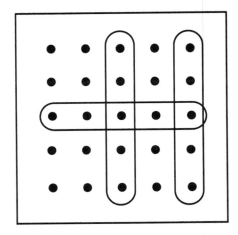

4. When the geoboard and dot paper are compared, the team gets five points for a match. Team members then switch jobs to continue.

5. At the end of a specified time, the class may compare scores and discuss why they were successful or not. The discussion may include clear or unclear directions, too difficult a design, or misunderstanding of terms, such as diagonal, perpendicular, intersecting, or parallel.

Variation: This activity could also be done using pattern blocks behind a barrier.

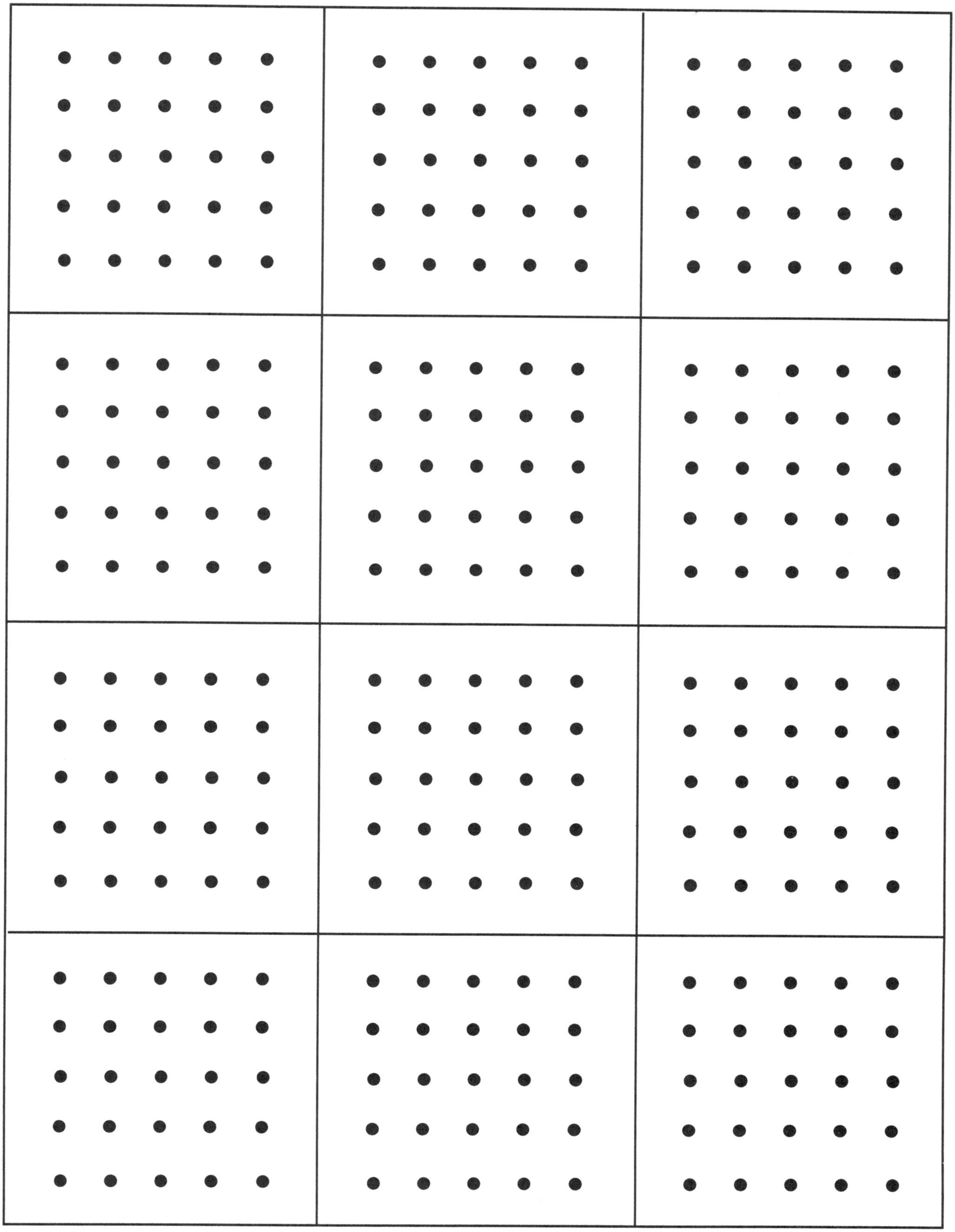

Geometric Mystery

Objective: to recognize geometric shapes by their properties

Materials needed: game cards

number of players: two players

Directions:

1. Shuffle the cards and keep them face down in a pile. The first player chooses a card from the pile but cannot look at the card. This player holds the card on his or her forehead for the other player to see.

2. The player must provide clues to help the first player identify the mystery geometric shape.
   For example: The first player shows a rectangle. The other player describes it using the following attributes and/or characteristics:
   - it has four right angles
   - opposite sides are parallel
   - it is a quadrilateral
   - the opposite sides are congruent

3. If the first player correctly identifies the shape, he or she earns a point and returns the card to the bottom of the pile.

4. If the first player does not correctly identify the shape, the card is returned to the bottom of the pile and no points are earned.

5. Play moves to the other player.

6. The winner is the player with the most points when time is called by the teacher.

trapezoid	rectangle	scalene triangle
isosceles triangle	equilateral triangle	square
rhombus	regular hexagon	regular octagon
acute triangle	right triangle	obtuse triangle

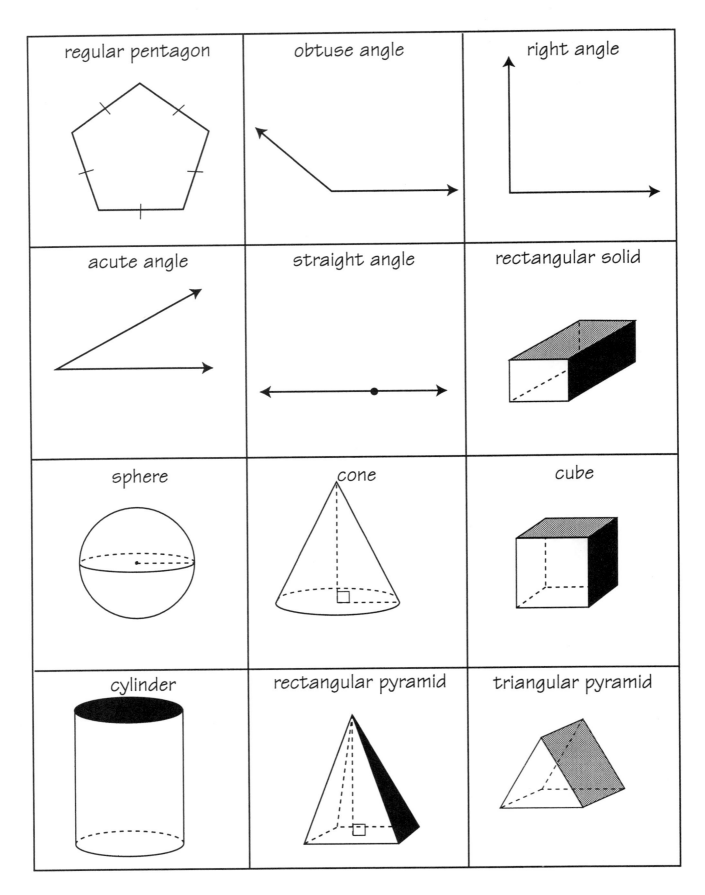

Math Monopoly

Objective: to provide practice calculating percents and using calculators

Materials needed:
paper and pencil
pair of dice or spinner with 1, 2, 3, 4, 5, 6
calculator
cards with the numbers 2, 3, 5, 7, 11 (see p.19)

number of players: two players

Directions:

1. Read all directions first. The game can go on for a long time, so you may want to set a time limit before beginning.

2. Take turns rolling the dice or spinning the spinner twice. Use the two numbers you roll or spin to form a two-digit number that represents the amount of money you have won on this turn. For example, if you roll a 2 and a 4, you may choose to win $24 or $42.

3. However, you must pay taxes on the number you choose! To calculate your tax, use the following tax schedule.
   TAX SCHEDULE
   $11 - $26   pay   15% of your number
   $31 - $46   pay   25% of your number
   $51 - $66   pay   35% of your number
   Remember: to calculate percent, convert the percent to a decimal fraction (15% becomes 0.15) and multiply the decimal fraction times the dollar amount you chose.

4. Calculate your tax and subtract this amount from your number. Round off after each operation. All dollar and cents amounts are rounded off to the nearest dollar. All 50¢ amounts are rounded up to the nearest dollar. (For example, $21.32 becomes $21.00; $24.85 becomes $25; and $28.50 becomes $29.00.)

   For example: you roll or spin a 2 and a 5. You choose $25. The tax on $25 rounds to $4.00 ($25 x 0.15 = $3.75; round $3.75 to the nearest dollar: $4.00). Subtract the tax from the original amount to determine how much you have won.

   First Roll $25   15% tax of $3.75 rounds to $4.00
                 -$4
                 $21 won

5. Keep a neat record of your accounts. See the section on Record Keeping.

PRIMES

6. After the first round of play, you can buy a prime number whenever you have enough money. The prime numbers you can buy are 2, 3, 5, 7, or 11. The prime numbers cost $500 times their reciprocal (For example, 7 costs $\frac{1}{7}$ x $500 or $71.43, which rounds off to $71.) (Remember: a reciprocal is the number that results when 1 is divided by a given number. The reciprocal of 5 is $\frac{1}{5}$, the reciprocal of 36 is $\frac{1}{36}$.)

7. If anyone chooses to use a number divisible by a prime that is owned by another player, he or she must pay a commission of 50% of the winnings after taxes. (Remember, divisible means that the number can be divided evenly by the prime, with no remainder.)

8. If you need cash, you can sell your prime number back for its cost minus 10% interest.

BONUS

9. If you roll a prime greater than 20, you get a 20% **bonus** before taxes. But you still must pay taxes on the new number. For example, for the number 31, you would get a $6 bonus and win $37. The tax would be 25% of $37, or $9.

Second Roll	$31	20% prime bonus of $6.20 rounds to $6
	+$6	bonus
	$37	
	-$9	25% tax of $9.25 rounds to $9
	$28	won

RECORD KEEPING

10. All players must keep a neat recording of their accounts.

11. Anyone may check another player's accounts before the turn passes and may charge a $10 fee for correcting any errors found.

Record in the following manner:

First Roll	$25	15% tax of $3.75 rounds to $4.00
	-$4	
	$21	won
Second Roll	$31	20% prime bonus of $6.20 rounds to $6
	$6	bonus
	$37	
	-$9	25% tax on $9.25 rounds to $9
	$28	won

Total so far: $21 + $28 or $49

12. The winner is the player to accumulate $500 or more before any other player.

Math Scavenger Hunt

Objective: to increase student awareness of real-world mathematics

Materials needed: "Math Scavenger Hunt" sheet

Number of players: whole class

Directions:

1. Divide the class into teams of four players each.

2. Teams read the "Math Scavenger Hunt" sheet and decide how they are going to divide the responsibilities.

3. Each item is worth one point. The team earns one extra point for every summary submitted with their findings. The summary must identify the mathematical concept and explain how it is used in real life.

4. Depending on your available resources, you may wish to assign this for a two-day period as a take-home assignment.

Math Scavenger Hunt

1. a ratio used in a magazine or the newspaper
2. a picture of a fraction
3. a sale with 20% off
4. a centimeter ruler
5. a picture of a car selling for more than $15,000
6. a rectangular prism
7. a cylinder
8. a cone
9. a pyramid
10. a line graph
11. a pie graph
12. a pictograph
13. a bar graph
14. a scattergram
15. a stem and leaf plot
16. a histogram
17. a pattern in material, glued on paper, and extended
18. an advertisement for a spreadsheet program
19. a math-related children's picture book
20. a postage scale
21. a parallelogram
22. a rhombus
23. a trapezoid
24. a pattern that tessellates
25. misleading statistics in an advertisement
26. parallel lines
27. perpendicular lines
28. lottery odds

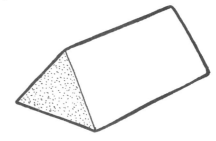

Metric Match

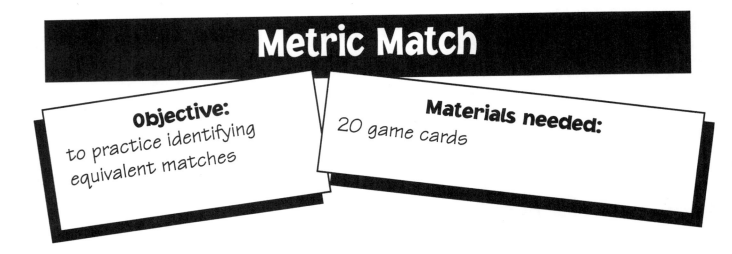

Objective: to practice identifying equivalent matches

Materials needed: 20 game cards

number of players: two to four players

Directions:

1. Shuffle the cards and lay them face down in rows. This game is like the game of concentration in which you have to remember where you saw the other equivalent fraction.

2. The first player turns over two cards trying to make a match. If the first player makes a match, he or she gets another turn. If the cards do not match, the cards are turned back over and the second player takes a turn.

3. The winner is the player with the most matches when all the cards are matched.

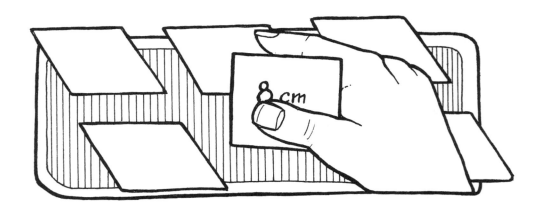

13mm	1.3cm	5cm	50mm
80mm	8cm	4.7cm	47mm
8cm	0.08m	13.9cm	139mm
59mm	5.9cm	1.7cm	17mm
50mm	5cm	45mm	4.5cm

Mystery Proportion

Objective: to solve for the unknown in a proportion

Materials needed: playing cards (cut out and laminated) paper and pencil calculator

number of players: two players

Directions:

1. One player shuffles the cards and places them face down.

2. The second player chooses three cards for the first player to create a proportion problem. For example, if 4, 15, and 36 are chosen, the first player may create the following proportion:

$$\frac{4}{15} = \frac{36}{}$$

3. The first player may choose any method to solve the proportion; however, that player must explain to the other player how he or she arrived at the answer.

4. The second player may use a calculator to verify the answer with cross products.

5. If the solution is correct, the first player scores a point and the roles are reversed. The first player then chooses three cards for the second player to create a proportion problem.

6. After 20 minutes (or longer), the player with the most points wins.

2	3	4	5
6	8	9	10
12	15	16	18
20	24	30	32
36	40		

Name That Symbol

Objective: to identify mathematical symbols

Materials needed: 24 symbol cards

number of players: whole class

Directions:

1. Divide the class into two teams. Write each team member's name on an index card and shuffle the two piles separately. Select one card from each pile. The two students chosen should go to the board.

2. Hold up a symbol card. The first student to correctly identify and legibly write the expression on the board wins a point for his or her team.

3. Choose two more name cards and call two new students to the board, one from each team. Again show the students at the board a symbol card. Each student tries to be the first to write the expression on the board.

4. The team with the most points at the end of ten turns is the winner.

Answer Key

$0.\overline{4}$	a repeating decimal	>	greater than
kg	kilogram	90°	90 degrees
//	parallel lines	GCF	greatest common factor
∟	right angle	LCD	least common denominator
7!	seven factorial	$\sqrt{}$	square root
1:5	the ratio of 1 to 5	$\overline{AB}$	segment AB
∴	therefore	<A	angle A
cm	centimeter	|x|	absolute value of x
≈	approximately	kilo	1,000
π	pi	m^3	m cubed
L	liter	πr^2	area of a circle
⊥	perpendicular	≠	is not equal to

$0.\overline{4}$	kg
∥	⌐
7!	1:5
∴	cm

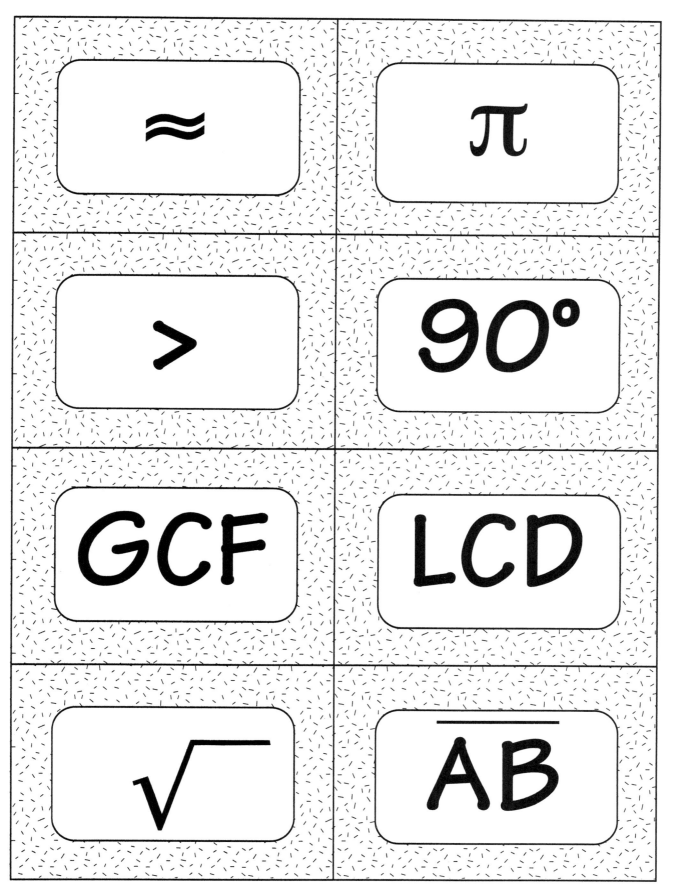

∡A	|x|
kilo	m³
πr²	L
⊥	≠

Number Mystery

objective: to determine the mystery number by asking clues

Materials needed: numbers written on a piece of paper, tape

number of players: whole group

Directions:

1. Each student tears a sheet of paper in half.

2. Each student writes a number on his or her sheet of paper. Students may not show their number to any other student.

3. Students tape their number on their neighbor's back, keeping the number a mystery.

4. Students are then asked to discover the number on their back by asking "yes" or "no" questions. They may only ask one question from each person.

   Examples: "Am I a prime number? Am I divisible by 9?"

5. When a student discovers his or her number, he or she becomes an advisor to help other students discover the number on their backs.

6. When you see that most students have solved the mystery of their number, ask students with the remaining unsolved numbers to stand in the front of the class. The entire class provides clues to help each remaining student discover his or her number.

Order War

Objective: to provide practice evaluating numerical expressions using the order of operation

Materials needed:
20 game cards
calculator

number of players: two players

Directions:

1. Shuffle the cards and divide them evenly between two players.

2. To begin, each player simultaneously turns over one card from his or her pile and evaluates the numerical expression.

3. The player with the greatest value wins the two cards.

4. If the two players have the same value, they turn over two more cards, evaluate the expression, and determine who has the greatest value. The player with the greatest value keeps all the cards.

5. Students may use calculators to verify their answers or to settle a challenge.

6. The winner is the player with the most cards at the end of play.

4x5+6	24-4x2	4x6+3x3	3^2+3x11
3x4-2x3	8÷2+7	5-32÷16	7^2-8x3
8x3+6	4^2+2x3	20-3^2	4x6-3x3
3+7+6x2	5+54÷6	2+3^2-8÷2	16+63÷9
5^2+6÷3-2	6x3-14÷7	48÷4+1	12+3÷3+1

Percent Match

Objective: to practice identifying the decimal and/or fraction equivalents for frequently used percents

Materials needed: one playing board for every member of the class

Number of players: entire class or small groups

Directions:

1. Provide each player with a playing card. Each player is to choose 16 percent values and write them in any of the smaller corner boxes until all are filled. The teacher or one student is to be the caller.

2. The caller reads one fraction or decimal equivalent at a time. Each player records the fraction or decimal equivalent in the larger box.

3. The winner is the first player to identify correctly all the fractions or decimal equivalents in any row, column, or four-square diagonal.

Fractions and decimals to be called out to the players:

1/3	1/2	5/8
2/3	2.25	1/4
1/8	4/5	0.0125
1.25	1/5	3/5
0.005	.75	0.025
.125	1	2
7/8	1/10	
0.0025	3/8	

Choose 16 percent numbers below and place one in each top corner.

33⅓%   66⅔%   12.5%   ½%   125%   87.5%   ¼%
50%   225%   20%   80% 75%   100%   10%
37.5% 62.5% 25%   200% 60%   1.25%   2.25%

Positive or Negative Numbers

Objective: to practice multiplying integers

Materials needed: deck of cards for each pair of players with the face cards taken out
paper to keep track of points
calculator

number of players: two players

Directions:

1. The first student turns over two cards. The black cards are positive numbers, and the red cards are negative numbers. The second player verifies the product using a calculator. The score is recorded on the paper.

2. The second player turns over two cards, multiplies the two numbers, and keeps track of the score. If the student turns over two black cards, or two red cards, the product is a positive number. If the cards turned over are one black and one red, the product is a negative number.

3. Students keep adding their scores of positive and negative numbers. When all the cards have been used, the play ends.

4. The student with the highest score wins.

5. Some students may need a visual of a number line in working with negative numbers.

Rational Madness

Objective: to practice multiplying rational numbers

Materials needed: 20 game cards, calculator

number of players: two players

Directions:

1. Shuffle the cards and divide them evenly face down between two players.

2. Both players turn their top card face up simultaneously.

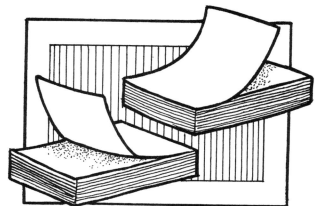

3. The first player to state the product of the two cards correctly in simplest form wins both cards.

   For example, player one turns over $3/2$ and player two turns over $-1/2$. Both players mentally compute the product of $(3/2)(-1/2)$. The first player to give the product takes the two cards.

4. The other player can challenge the answer and use a fraction calculator to verify the answer. If that player proves the answer wrong, the two cards are awarded to him or her.

5. The winner is the player with the most cards when all cards have been used.

$-\dfrac{2}{3}$	$\dfrac{7}{8}$	-2	-3
-4	5	6	$-\dfrac{3}{4}$
$\dfrac{5}{12}$	$-\dfrac{3}{20}$	$\dfrac{4}{5}$	$\dfrac{3}{2}$
0	1	$-\dfrac{1}{2}$	$-\dfrac{5}{6}$
$\dfrac{1}{6}$	$\dfrac{7}{12}$	$\dfrac{11}{12}$	$-\dfrac{5}{8}$

Shapes, Shapes, Shapes

Objective: to practice naming shapes on a gameboard

Materials needed:
a copy of the gameboard
markers
answer sheet to verify shapes
number cubes

number of players: two to four players

Directions:

1. Two to four players take turns throwing the number cubes and moving each marker the specified number of blocks.

2. If a player names the shape and another player challenges, they can check the answer sheet. If the player is wrong, he or she loses the next turn.

3. The goal is to get all the way to the end, turn around, and go all the way back to the start. If a player is only two blocks away from the end and rolls a five, he or she must use two blocks to get to the end and three blocks to head back to the start.

4. Players must roll the exact number to land back on "Start" to win the game.

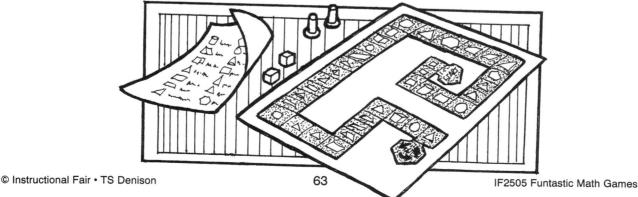

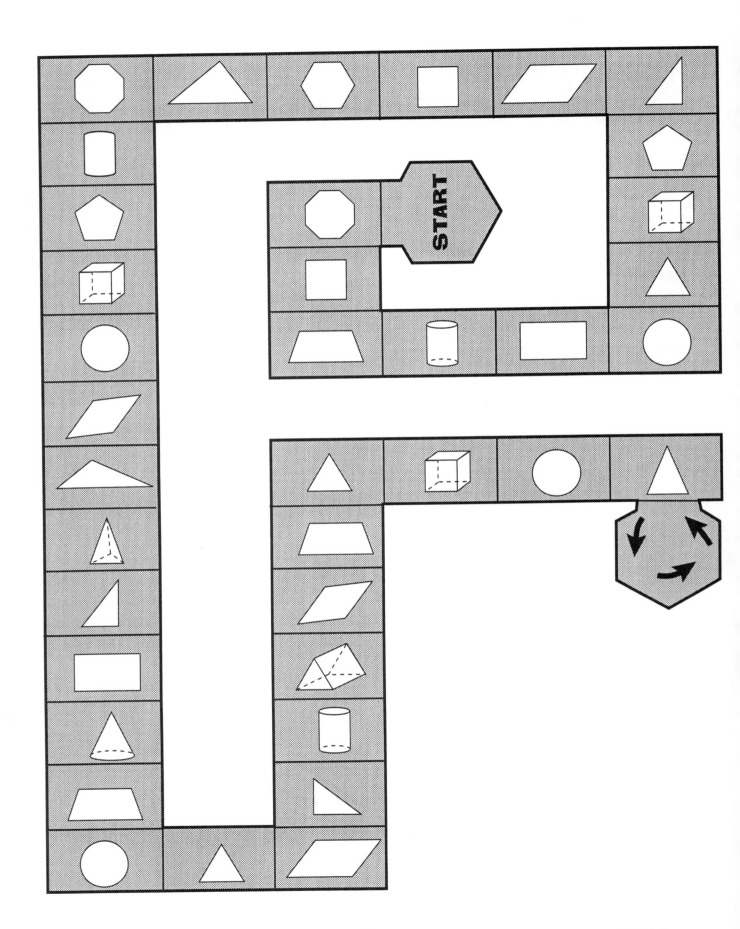

Answer Sheet

- cylinder
- triangular prism
- cube
- triangular pyramid
- parallelogram
- trapezoid
- rhombus
- scalene triangle
- right triangle

- square
- octagon
- pentagon
- circle
- cone
- rectangle
- isosceles triangle
- equilateral triangle
- hexagon

Spin-Roll Expressions

Objective: to evaluate simple algebraic expressions

Materials needed:
spinner with one side positive one side negative
a number cube
45 playing cards (copy on cardstock and then laminate)

number of players: three to four players

Directions:

1. Each player rolls the number cube. The player with the lowest number is the dealer.

2. The dealer passes out all cards and then rolls the number cube. The number on the number cube is the value of the variable.

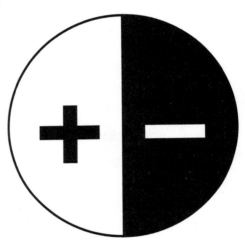

3. The dealer spins the spinner to determine if the value from the number cube is positive or negative.

4. The player to the right of the dealer places a card on the table and explains its value. Play proceeds to the next player until all players have evaluated their expression and placed their card on the table. The player with the greatest value for that round wins all of the cards.

5. The dealer rolls the number cube and spins the spinner to determine the next variable value. Play continues until all cards are claimed.

6. The player with the most cards at the end is the winner.

x^2	$2x$	$2x+1$
$3x$	$3x+1$	x^3
$3x-1$	$2x-1$	$3x+2$

$5x + 1$	$5x$	$5x - 1$
$x^2 - 1$	$x^2 + 1$	$x^3 + 1$
$x^3 - 1$	$3x - 2$	$x^3 - 2$

$x^3 + 2$	$x^2 - 2$	$x^2 + 2$
$3x - 5$	$3x + 5$	$5x - 3$
$5x + 3$	$2x^2$	$2x^2 + 1$

$2x^2 - 1$	$2x^3 + 1$	$2x^3 - 1$
$4x$	$4x + 1$	$4x - 1$
$4x + 2$	$4x - 2$	$3x + 3$

$2x+3$	$2x-3$	$2x+5$
$2x-5$	$5x+2$	$5x-2$
$2x^3$	$2x^3-1$	$2x^3+1$

Square T²

Objective: to provide practice recognizing perfect square roots

Materials needed: 20 playing cards cut out and laminated a tic-tac-toe frame (Students can draw their own tic-tac-toe frame.)

number of players: two to four players or the entire class

Directions:

1. Mix the 20 playing cards and place them face down.
2. Each player draws a tic-tac-toe frame on a clean sheet of paper.
3. Each player may choose to place any nine of the following numbers in the tic-tac-toe frame: 1, 2, 3, 4, 5, 6, 7, 8, 9, 10, 11, 12, 13, 14, 15, 16, 17, 20, 25, 30. Each number may only be used once.
4. The teacher or the caller chooses one card from the deck and reads it aloud to the players. The teacher or caller should list the numbers read to check the winner's card. The card is placed on the bottom of the stack.
5. Any player with the correct square root value on his or her tic-tac-toe frame crosses out that square.
6. The teacher or the caller takes another card from the deck and reads it aloud. Players mark their cards again if they have the correct square root value.
7. The winner is the first player to mark three Xs in a vertical, horizontal, or diagonal row.

$\sqrt{1}$	$\sqrt{4}$	$\sqrt{9}$	$\sqrt{16}$
$\sqrt{25}$	$\sqrt{36}$	$\sqrt{49}$	$\sqrt{64}$
$\sqrt{81}$	$\sqrt{100}$	$\sqrt{121}$	$\sqrt{144}$
$\sqrt{169}$	$\sqrt{196}$	$\sqrt{225}$	$\sqrt{256}$
$\sqrt{289}$	$\sqrt{400}$	$\sqrt{625}$	$\sqrt{900}$

Stars and Pounds

Objective: to find as many possible answers for a literal equation as possible

Materials needed:
16 game cards
stop watch
team sheet (blank sheet of paper)

number of players: teams of two players

Directions:

1. Divide the class into pairs. The teacher chooses a card at random and writes the literal equation on the overhead or the board. Each pair of students will work together to write as many solutions for the given equation in two minutes on their team sheet.

2. Students may use any number set in their solutions.

3. Make a list on the board as each pair of players shares one solution. If the players do not have an original solution, they pass. Continue writing all the possibilities until the list is complete.

4. The team of students with the greatest number of solutions is the winner for that round.

Sample Literal Equation Solutions

■ + ★ = 12

Possible solutions:

0 + 12	1 + 11	2 + 10	3 + 9	4 + 8	5 + 7
6 + 6	7 + 5	8 + 4	9 + 3	10 + 2	11 + 1
12 + 0	-1 + 13	-2 + 14	-3 + 15	-4 + 16	. . . .

■ + ★ = 12	# + ❀ = 16
♥ + ✳ = 4	➤ - 9 = 11
▲ + ✖ = 10	✺ + ☽ = 21
A + B = 7	23 - 🍒 = 14

□ − ★ = 5	○ − ✦ = 13
✦ − ♠ = 18	✠ + 6 = 25
♦ + ✧ = 0	❀ − ❀ = 0
▮ + ▼ = −4	⦁ − ● = −8

Strategic Plotting

Objective: to provide practice plotting points on a coordinate plane

Materials needed: graph paper

Number of players: two players

Directions:

1. This game is played like tic-tac-toe. The object is to get five Xs or five Os in a row (horizontally, vertically, diagonally) by plotting points on graph paper and to prevent an opponent from plotting five points in a row.

2. Players decide who will be the X and who will be the Y.

3. Taking turns, each player states, writes, and plots a coordinate pair. The X player marks an X on the coordinate position. The O player marks an O on the coordinate position.

4. Once the coordinate pair is stated and written, it cannot be changed. If a player states a coordinate pair that has already been used, he or she loses a turn.

5. The winner is the first player to get five in a row.

Player One

X	Y

Player Two

X	Y

Stem-and-Leaf Logic

Objective: to use logic to place numbers in a stem-and-leaf plot to achieve the sum closest to 100

Materials needed: one playing sheet, two number cubes

number of players: two players

Directions:

1. The two players share the same game sheet but must use different colors to record their numbers.

2. Each player rolls the dice and must decide how to place the digits in the stem-and-leaf plot. The dice may be added, subtracted, or combined to form a double-digit number. For example, if a player rolls 4 and 1, that player must decide if he or she is going to use 3, 5, 14, or 41 in the stem-and-leaf plot.

3. Each player takes a turn rolling the dice.

4. The winner of the game is the player whose sum is closest to 100 after five turns.

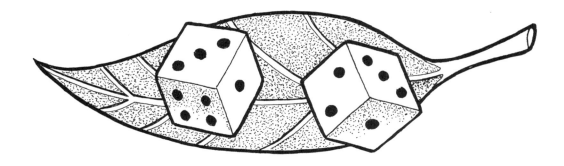

Example: If you roll 3 and 6, you must decide if you want to use 3, 9, 36, or 63. The tens place is the stem, and the leaf is the ones place.

Stem	Leaf	Player 1 Sum	Player 2 Sum
0			
1			
2			
3			
4			
5			
6			
7			
8			
9			

Sweet Success

Objective: to provide practice in estimating products

Materials needed: "Wheel of Fortune" spinners, a calculator, Hershey kisses

number of players: two to four players

Directions:

1. Each player spins one of the "Wheel of Fortune" spinners. The player spinning the largest number begins the game by going first. The player spinning the second largest number will go next.

2. Pencils and paper may not be used throughout the game.

3. The first player begins the game by spinning both spinners and estimating the product of the two numbers.

4. The same player then selects the range of numbers which contains the estimate from the "Range of Products" list.

5. The player then checks his or her answer using the calculator that has been provided.

6. If the player's answer is correct, the player may treat himself or herself to a Hershey kiss. If the player's answer is not correct, the player is not awarded any candy kisses.

7. Each player follows in turn.

8. The winner is the player who has experienced the greatest number of successes and obtained the most Hershey kisses at the end of ten rounds or at the end of a designated time period.

Sweet Success

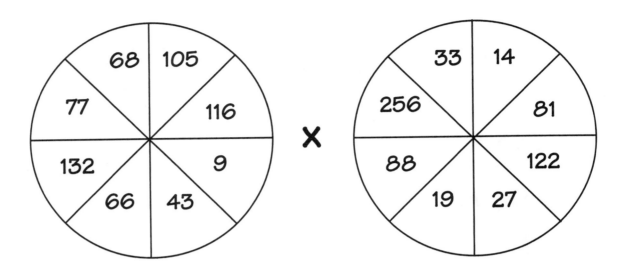

Range of Products

(1.) 100 - 999

(2.) 1,000 - 1,999

(3.) 2,000 - 2,999

(4.) 3,000 - 3,999

(5.) 4,000 - 4,999

(6.) 5,000 - 5,999

(7.) 6,000 - 6,999

(8.) 7,000 - 7,999

(9.) 8,000 - 8,999

(10.) 9,000 - 99,999

Twins

Objective: to practice translating sentences into algebraic equations

Materials needed:
18 sentence game cards
18 equation game cards

number of players: whole class

Directions:

1. Shuffle and distribute one card to each player making certain there is a twin for each sentence.

2. After looking at their cards, players will search for their twin or match. Every player who finds his or her twin is a winner.

One number is six times another.	x = 6y	The sum of three consecutive whole numbers is 15.	x + (x + 1) + (x + 2) = 15
One number is three less than another number.	x = y - 3	One number is five more than another number.	x = y + 5
Seven more than twice a number is 15.	2x + 7 = 15	Double a number increased by 8 is 24	2x + 8 = 24
Triple a number decreased by 8 is 19.	3x - 8 = 19	Forty increased by a number is four times the number.	40 + x = 4x
Double a number increased by 4 is 20.	2x + 4 = 20		

Seventeen is four more than x.	17 = x + 4	Thirty-six is half of x.	$36 = \dfrac{x}{2}$
X decreased by seven is 23.	x - 7 = 23	Two more than twice x is 18.	2x + 2 = 18
Eighty-three is 12 less than x.	83 = x - 12	The quotient of x and three is seven.	$\dfrac{x}{3} = 7$
The product of seven and x is 42.	7x = 42	The sum of x and 35 is 83.	x + 35 = 83
The difference of x and 17 is 37.	x - 17 = 37		

Westward Go!

Objective: to provide practice for rounding numbers to the greatest place value

Materials needed:
"States" gameboard (Reproduce pages 91 and 92 and tape them together.)
a token for each player
36 game cards
"Chance Cards"

number of players: two to four players

Directions:

1. Shuffle the 36 game cards and the "Chance Cards" together and place the deck face up in the middle of the playing area.
2. All players begin by placing their tokens on the cars on the gameboard.
3. The player whose first name is closest to the beginning of the alphabet begins the game by drawing a card and rounding the number on the card to the greatest place value. For example, 1,555 rounded to the greatest place value is 2,000.
4. If the player's answer is correct and that number is located in the first state on the gameboard, the player moves his or her token to that state. If the number is not located in the first state, the player does not move the token.
5. Each player follows in turn, drawing a card, rounding the number on the card to the greatest place value, and moving the token to the first state, if possible.
6. The correct answer must always be in the state directly ahead on the gameboard for the player to move on any turn.
7. If a "Chance Card" is drawn from the pile, the player must follow the directions on the card.
8. The winner of the game is the first player to reach the last state or the player who is ahead at the end of the designated time period.

Chance Cards

Lose your next turn.	Move ahead to the next state that contains 2,000.	Go back one state.
Move ahead to the next state.	Move back two states.	Gain an extra turn.
Move back to the car.	Move ahead to the next state that contains 3,000.	Move back two states.
Move back to the smallest state.	Move back to the largest state.	

Game Cards

1,000	1,010	1,013
1,122	1,225	1,327
1,555	1,897	1,999
1,498	1,873	1,455

2,133	2,671	2,390
3,456	6,651	4,200
5,300	6,499	6,944
8,571	8,322	7,813

8,199	5,455	2,973
8,990	9,382	7,501
6,776	9,811	8,700
2,835	6,164	2,222

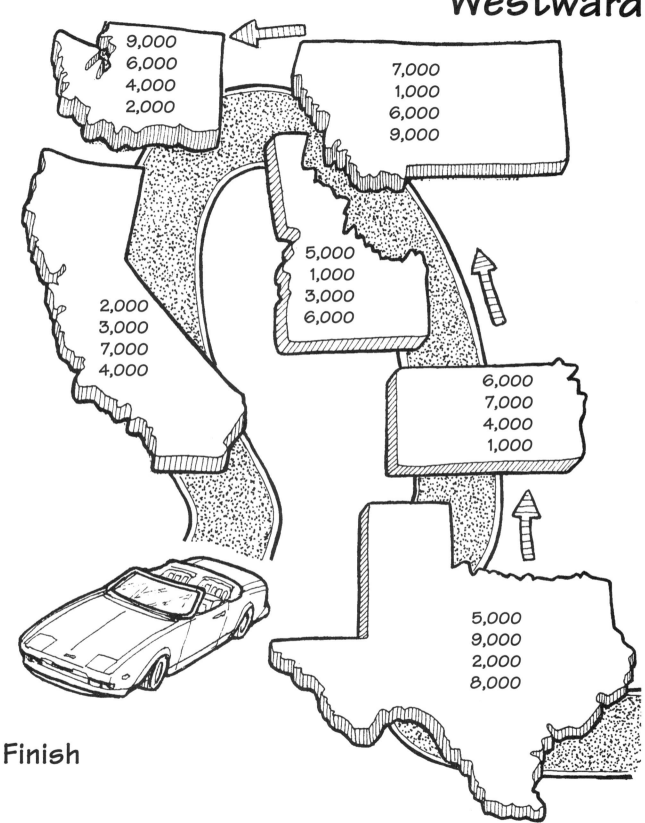

Go

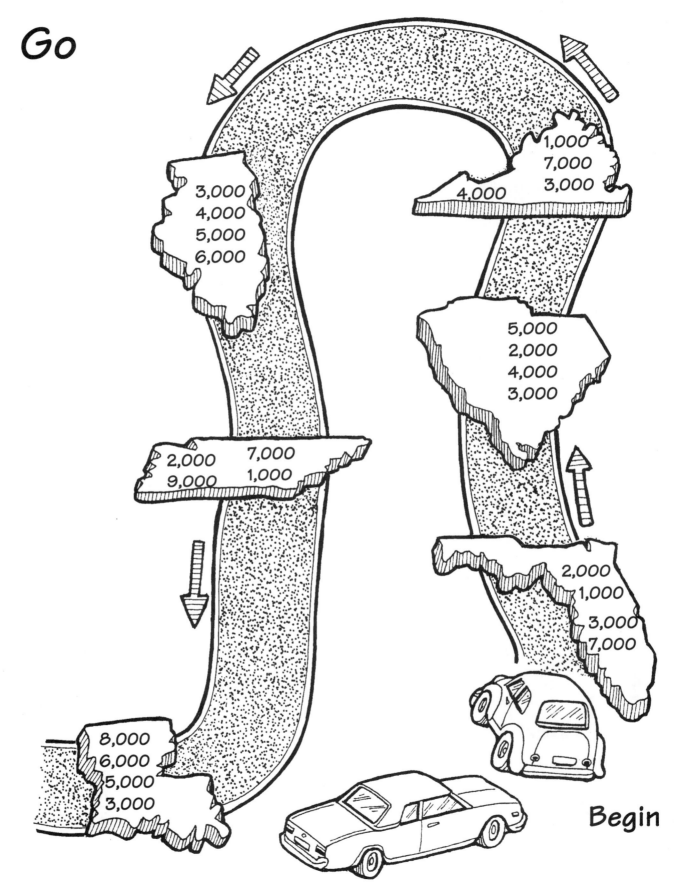

Begin